D1227183

An Elementary Guide to

RELIABILITY

FOURTH EDITION

Pergamon Titles of Related Interest

BADEN FULLER
Microwaves, 3rd Edition

BROOK & HANSTEAD
Reliability in Non-Destructive Testing

GANDHI
Microwave Engineering and Applications

HAMMOND
Electromagnetism for Engineers, 3rd Edition

HARRISON
Structural Analysis and Design, 2nd Edition

HAYWOOD
Analysis of Engineering Cycles, 3rd Edition

HOBBS & DOLING
Planning for Engineers and Surveyors

HOLLAND
Integrated Circuits and Microprocessors
Microcomputers and Their Interfacing
Microprocessors and Their Operating Systems

MOTTERSHEAD
Modern Practice in Stress and Vibration Analysis

RAO
The Finite Element Method in Engineering, 2nd Edition

Pergamon Related Journals

Computers and Electrical Engineering
Computing Systems in Engineering
Corrosion Science
Electrochimica Acta
Engineering Fracture Mechanics
Expert Systems with Applications
Fatigue and Fracture of Engineering Materials and Structures
International Journal of Applied Engineering Education
International Journal of Machine Tool Design and Research
Journal of the Mechanics and Physics of Solids
Mechatronics
Microelectronics and Reliability
Solid State Electronics

(free specimen copy gladly sent on request)

An Elementary Guide to
RELIABILITY
FOURTH EDITION

by

G. W. A. DUMMER
MBE, CEng, FIEE, FIEEE

and

R. C. WINTON
MBE, BSc(Eng), ACGI, CEng, FIEE

PERGAMON PRESS
Member of Maxwell Macmillan Pergamon Publishing Corporation

OXFORD · NEW YORK · BEIJING · FRANKFURT
SÃO PAULO · SYDNEY · TOKYO · TORONTO

U.K.	Pergamon Press plc, Headington Hill Hall, Oxford OX3 OBW, England
U.S.A.	Pergamon Press, Inc., Maxwell House, Fairview Park, Elmsford, New York 10523, U.S.A.
PEOPLE'S REPUBLIC OF CHINA	Pergamon Press, Room 4037, Qianmen Hotel, Beijing, People's Republic of China
FEDERAL REPUBLIC OF GERMANY	Pergamon Press GmbH, Hammerweg 6, D-6242 Kronberg, Federal Republic of Germany
BRAZIL	Pergamon Editora Ltda, Rua Eça de Queiros, 346, CEP 04011, Paraiso, São Paulo, Brazil
AUSTRALIA	Pergamon Press Australia Pty Ltd., P.O. Box 544, Potts Point, N.S.W. 2011, Australia
JAPAN	Pergamon Press, 5th Floor, Matsuoka Central Building, 1-7-1 Nishishinjuku, Shinjuku-ku, Tokyo 160, Japan
CANADA	Pergamon Press Canada Ltd., Suite No. 271, 253 College Street, Toronto, Ontario, Canada M5T 1R5

Contents

Introduction   vii

1. The Importance of Reliability   1

2. Definitions of Reliability   4

3. Some Simple Statistics   9
   Probability, product law, confidence levels, the language of
   reliability statisticians

4. How Reliability is Calculated   17
   MTBF, physics of failure, failure rates, derating, life expectancy,
   redundancy, quality control, stages in equipment design and
   production

5. The Effect of Operating Conditions and Environments   29
   Temperature, humidity, etc.

6. Mechanical Reliability   32
   General mechanical failures, guidelines for designers

7. Installation and Operability   35
   Ergonomics

8. Maintainability   40
   Availability, test equipment, MTTR

9. Reporting Failures   46
   Aims, feedback

10. The Cost of Reliability   48
    Total costs

11. Some Useful Reliability Definitions   53

References for Further Reading   59

v

Introduction

An Elementary Guide to Reliability explains in simple, largely non-technical language what is meant by reliability and the various factors which make an equipment or machine reliable. It deals with basic considerations which apply equally to electrical, electronic, or mechanical designs, even though most examples are drawn from electronics, where reliability is more widely accepted and which is the authors' main field of experience.

The Guide has been written to provide an introduction to reliability for those without any previous knowledge of it. However, even those who are already familiar with some aspects of reliability should find the book of interest, since it covers all basic facets of reliability.

Those who need to obtain some knowledge of reliability in the course of their studies, or of their work, can use this Guide to acquire the basic concepts, on which more detailed and more technical knowledge can subsequently be based. In particular the Guide will serve as a textbook for teachers and students concerned with the following courses:

City and Guilds of London Institute
Course 275—Industrial Measurement and Control Technicians Certificate, Section 275–3–23 Logic, Data Handling and Digital Computer Control Systems.
Course 743—Certificate in Quality Control.
Course 803—(This course is designed for overseas students)—Electrical Engineering Technicians Certificate, Section 803–3–21 Advanced Electrical Technology.
Business and Technician Education Council.
        College devised course units which include the basic concepts of reliability or of quality assurance in the syllabus.

Those, whether technical or laymen, who are interested in learning something about reliability in an age when it is assuming more and more importance, will find the background they need in these pages.

Many books on reliability deal mainly with particular aspects in specialized language, which is usually beyond the grasp even of the technical man without previous knowledge of the subject. It is this gap which the Elementary Guide is intended to fill.

G. W. A. DUMMER
R. C. WINTON

Acknowledgements

The authors would like to thank Professor K. B. Misra of the Reliability Engineering Centre at the Indian Institute of Technology, Kharagpur, India, for his help in providing definitions in the chapter on the "Language of Reliability Statistics" and also for the section on "Some useful reliability definitions".

Acknowledgement is made to the valuable contributions and comments by Mr. M. Bedwell of the Combined Engineering Department, Coventry Polytechnic.

Acknowledgement is also made to Professor B. S. Dhillon of the University of Ottawa, for data based on his book *Mechanical Reliability: Theory, Models and Applications*, in the section on mechanical reliability.

CHAPTER 1

The Importance of Reliability

Electrical, electronic and mechanical equipment is used in a number of fields—in industry for the control of processes, in computers, in medical electronics, atomic energy, communications, navigation at sea and in the air and many other fields. It is essential that this equipment should operate reliably under all the conditions in which it is used. In the air navigation, military and atomic energy fields, for instance, failure could produce a dangerous situation. Very complicated systems, involving large numbers of separate units, such as space electronics, are coming into use more and more. These systems use an extremely large number of parts, and as each individual part is liable to failure, the overall reliability will decrease unless the reliability of each component part can be improved.

Suppose, for instance, it is known that one component out of half a million would break down every hour. Then an equipment using 100,000 of these components would break down at an average interval of 5 hours.

The requirement for reliability is different for each application. In the transatlantic cable service, for instance, the underwater amplifiers must operate for 20 years or so without failure, because the cost of raising the cable to repair a failure would be about £100,000 (since it would be necessary to send a cable ship to the location, find the failure under several miles of ocean, supply and install a new amplifier, lower the cable to the bottom again and return to port). Added to this is the loss of revenue while the cable is out of action, which might bring the total to £500,000, or more.

In Britain the Air Registration Board will only license aircraft to use a blind landing system if the fault rate for the total system is less than one in ten million.

In the case of military aircraft on a mission, or a missile flight, it is vital for the equipment to operate for the period of flight, or the mission and perhaps a battle might be lost. It has been estimated that unreliability costs the R.A.F. alone up to £100 million each year in spares and servicing costs. In the case of an equipment controlling a chemical plant or some complex industrial process, the cost of "shut-down" may be considerable in spoiled production and loss of output.

It is essential to "build-in" reliability by sound design and construction and to carry out enough tests to make sure that this has been done.

The "availability" or time an equipment is functioning correctly while in use depends both on reliability and on maintainability. Reliability is defined in detail in the next chapter, but may be said to be a measure of an equipment's ability to perform its functions consistently under given conditions. Maintainability is a measure of the speed with which loss of performance is detected, diagnosed and made good, and this is discussed in Chapter 7.

Reliability is of course a most important factor in the safety of an equipment, but it is by no means the only factor. A system or equipment can "perform its required function" (see definition on page 4) and yet be unsafe. There are well-documented major disasters caused not by mechanical or electronic breakdown but by human failure. Such human failures can be due to failures of operators, of maintenance staff, or of management.

Breakdowns and disasters can arise from operators' failure, accidental or deliberate, to observe laid down operating procedures and regulations. The results of accidental failure can be allowed for to some extent by anticipating, during the equipment design stage, what accidental operating mistakes might be made and introducing safeguards. However, it is impossible to anticipate deliberate failure to

FIG. 1.1. "It's only a random failure, Sir".

follow operating procedures, and it is essential to impress on operators and on maintenance staff, during their training, how vital it is that they never depart from laid-down procedures and regulations however safe it may appear to do so.

Figures 1.1, 1.2 and 1.3 illustrate in a lighthearted way the importance of reliability and some of its aspects.

FIG. 1.2. One test is worth 10,000 opinions.

FIG. 1.3. You are being flown by an automatic pilot.

CHAPTER 2

Definitions of Reliability

Because of the many differing operational requirements and varying environments, "reliability" means different things to different people. The generally accepted definition of reliability* is

> *"Reliability—the characteristic of an item expressed by the probability that it will perform a required function under stated conditions for a stated period of time".*

It will be noted that by definition reliability is a *probability* of success and this aspect is dealt with in more detail in Chapter 4.

General definitions are given below whilst fuller definitions are given in Chapter 11.

A "failure" is any inability of a part or equipment to carry out its specified function.

An "item" may be any part, sub-system, system or equipment which can be individually considered and separately tested.

An item can fail in many ways and these failures are classified as follows.

(a) Causes of failure:
   (i) *Misuse failure*
   Failures attributable to the application of stresses beyond the stated capabilities of the item.
   (ii) *Inherent weakness failure*
   Failures attributable to weakness inherent in the item itself when subjected to stresses within the stated capabilities of that item.
(b) Times of failure:
   (i) *Sudden failure*
   Failures that *could not* be anticipated by prior examination.
   (ii) *Gradual failure*
   Failures that *could* be anticipated by prior examination.

* B.S.I. (British Standards Institution) and I.E.C. (International) definitions.

(c) Degrees of failure:
  (i) *Partial failure*
  Failures resulting from deviations in characteristic(s) beyond specified limits *not* such as to cause complete lack of the required function.
  (ii) *Complete failure*
  Failures resulting from deviations in characteristic(s) beyond specified limits such as to cause complete lack of the required function. The limits referred to in this category are special limits for this purpose.
(d) Combinations of failures:
  (i) *Catastrophic*
  Failures which are both sudden and complete.
  (ii) *Degradation*
  Failures which are both gradual and partial.

An important criterion that the user or the maintenance engineer must know is how often the item breaks down and this is defined in two ways.

1. Mean time between failures—MTBF
  This applies to *repairable* items, and means that if an item fails, say, five times over a period of use totalling 1,000 hours, the mean (or average) time between failures would be 1,000 divided by five or 200 hours.
2. Mean time to failure—MTTF
  This applies to *non-repairable* items, and means the average time an item may be expected to function before failure. It is found by stressing a large number of the items in a specified way (e.g. by applying certain electrical, mechanical, heat or humidity conditions), and after a certain period dividing the length of the period by the number of failures during the period.

General relationship between terms

The diagram shown on page 6 gives the general relationship between variants of reliability characteristics—for example, the possible variants of failure rate, mean time between failures, mean time to failure and mean life.

Equipment failure pattern during life

When a new purchase is made, whether it is a motor car, radio set, washing machine or an aeroplane or tank, early failures may occur. These may be caused by manufacturing faults, design faults or misuse. The early failure rate may, therefore, be relatively high, but falls as

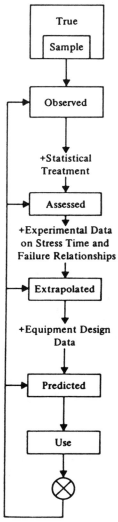

The value determinable only when the whole population has failed or has been subjected to stated stress conditions for a stated time.

The value relating to a sample subjected to stated stress conditions for a stated time.

Note. The sample may be equal to the whole population; in this case the observed value is equal to the true value.

A limiting value relating to a population of the confidence interval with a stated probability level based on observed data. This value may be the upper or lower limit of the confidence interval, as the case may be.

The value relating to different time and/or stress conditions.

The value relating to an item based for example on the failure rates of its parts.

The item may now be used or tested and its reliability characteristic values, subject to suitable variations for conditions applying to the use or tests, compared with those variants referred to above.

the weak parts are replaced. There is then a period during which the failure rate is lower and fairly constant, and finally the failure rate rises again as parts start to wear out. This is illustrated in Fig. 2.1, where the high rate of initial failures can be seen. Although the steady rate is often shown as a straight line, in practice it will be wavy and in good (reliable) equipments it may be a long time before the wear-out period is reached. The part of main interest is the constant failure rate period. The three parts are defined by the B.S.I. as follows:

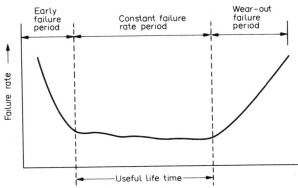

FIG. 2.1. Bath tub curve.

Early failure period

That early period, beginning at some stated time and during which the failure rate of some items is decreasing rapidly.

Constant failure rate period

That period during which failure occurs in some items at an approximately uniform rate.

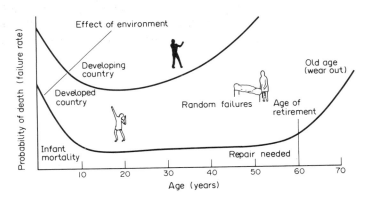

FIG. 2.2. Human life characteristics.

Wear-out failure period

That period during which the failure rate of some items is rapidly increasing due to deterioration processes.

Figure 2.2 compares the practical aspects of human life with the kind of curve illustrated in Fig. 2.1. It also illustrates the effect of environment on reliability (or life) which is discussed in Chapter 5.

CHAPTER 3

Some Simple Statistics

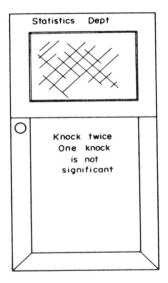

FIG. 3.1. Statistics Department.

Reliability is concerned with quantities, e.g. How many units are in use? How many parts? How many fail? What proportion is this to the total? Statistics are used in analysing these quantities. Statistics are also an essential tool in calculating probabilities of failures. Mathematically, "probability" can be expressed in three ways, all of which in practice mean the same thing.

1. As a percentage, e.g 99%
2. As odds, e.g. 99 : 1
3. As a "ratio", e.g. 0.99.

9

What most intending users want to know is the probability of survival of all the parts in an equipment, i.e. the reliability. If we start (time = 0) and see what the probability of failure is some time later (time = *t*), we find that practical experience of faults in equipments suggests some form of mathematical equation. This equation is based on what is known as the exponential law of reliability. This says that, for a given initial population, the rate of failure decreases with time only in so far as there are progressively fewer survivors left to fail.

The probability of no failures occurring in a given time can be expressed as follows:

$$R = e^{-\lambda t}$$

where *R* is the probability of no failures in time *t*,

     e is the base of the natural system of logarithms and equals 2.718,

     λ is the constant failure rate.

As   MTBF (mean time between failures) $= \dfrac{1}{\lambda}$

or $\lambda = \dfrac{1}{\text{MTBF}}$,

then   Reliability $= e^{-t/m}$

where *m* = MTBF,

     t = time.

How to calculate the MTBF is described fully in the next chapter, but it would be useful to look at some simple statistics in faults analysis. If an equipment is operating and failures occur, the number of failures can be recorded at the time at which they happen as shown in Table 3.1.

These observations can be plotted in the form of a graph from which more simple statistics can be discussed (see Fig. 3.2). Curve (1) records the actual failures as they occur with time. Curve (2) shows the cumulative total as time goes on.

Only seven observations were made in this simple example, but supposing a very large number of observations were made—for example in nuts and bolts, or transistors, or any small part where large quantities are involved. Then the *frequency distribution* of the measurements will be important.

It is generally convenient to group the measurements in the form of Table 3.2.

If this is plotted in the form of a curve, it will show the scatter of the measurements (see Fig. 3.3).

If the frequency of observation is plotted against the value of observations, a curve similar to that shown in Fig. 3.4 is obtained. This gives

TABLE 3.1

	Time (hours)	Failures	
	0	0	
	5	1	
	12	2	= 7 observations
	18	1	or measurements
	26	3	
	38	2	
	56	5	
Totals	56	14	

As 14 failures occur in 56 hours, the MTBF is $\frac{56}{14} = 4$ hours.

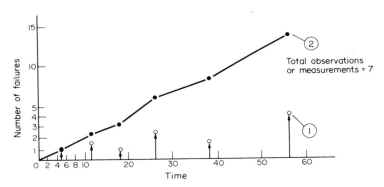

FIG. 3.2. Simple rate and cumulative graph.

TABLE 3.2

Measurement value	Number of times (or frequency) that this value occurs
1	1
2	3
6	4
22	7
31	9
34	8
39	4
43	3
49	2
64	1

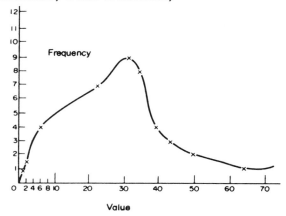

FIG. 3.3. Frequency curve showing scatter of measurements.

a graphical representation of a type of distribution known in statistics as "Normal" or "Gaussian". The curve is symmetrical about the centre line and the frequency at the centre line of symmetry is the most frequently occurring observation (or "Mode").

The "standard deviation" is a measure of the scatter of observations about the centre line and is usually represented by the Greek letter σ.

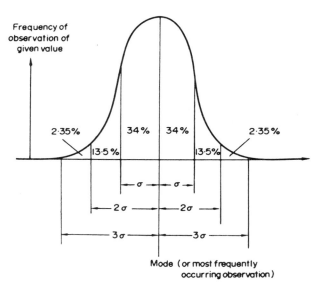

FIG. 3.4. Deviations of area from the mode.

If we consider the area under the curve in Fig. 3.4, the proportion of the total area which lies between $+\sigma$ and $-\sigma = 68\%$ or in other words 68% of the measurements may deviate from the centre line (mode) by amounts equal to less than $\pm\sigma$. Similarly,

95% of the measurements lie between $+2\sigma$ and -2σ
99.7% of the measurements lie between $+3\sigma$ and -3σ

so that a standard deviation of $\pm3\sigma$ is often used to cover well over 99% of all measurements.

The Normal or Gaussian distribution is used when the number of observations is large.

Another type of distribution encountered in reliability work is the "Poisson" and this is used to describe the situation when events occur at random but at a constant "rate", e.g. to individual failures of equipment with a known MTBF, which is much more important than the "Normal" in reliability work.

Other types of distribution, e.g. "Gamma" and "Weibull" are also of considerable interest to the reliability engineer, but these fall outside the scope of the book and full descriptions of them can be found in references 8 and 10 on page 59. Some more technical terms used by statisticians are given later in the section on "The language of reliability statistics".

The product law of reliability

It is obvious that the more parts there are in any machine or equipment the more risk there is of any one failing.

With only two parts such as a motor car engine and a motor car chassis, let us suppose that tests on six of each are made. If the six tests on the engine result in five engines passing the tests and one failing this could be simply represented as in Fig. 3.5.

As there are six chassis, suppose the tests on the chassis result in five passes again, and one failure, then this can be represented pictorially as shown in Fig. 3.6.

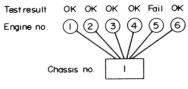

FIG. 3.5. One engine failure in six.

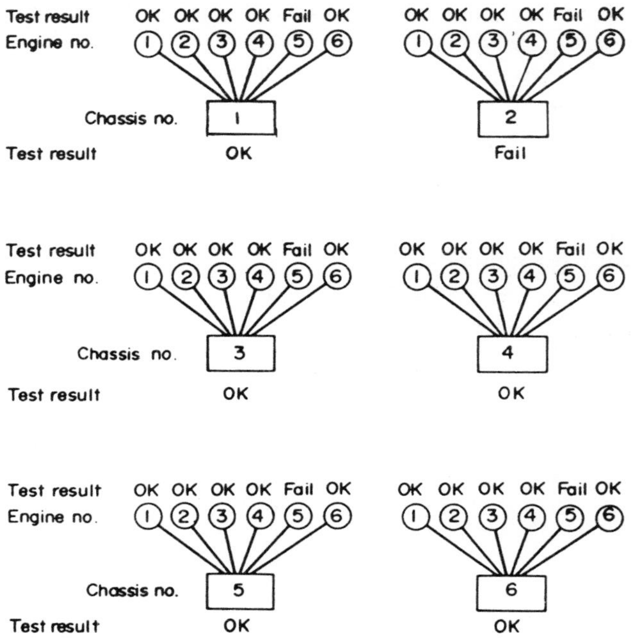

FIG. 3.6. Combinations of engine and chassis failure.

The reliability of each engine or each chassis is that five pass the test out of six or 5/6 = 0.833 or 83%. Engine No. 5 and chassis No. 2 fail and we can therefore indicate a failure against any combination including one of these (Fig. 3.6). There are, therefore, eleven failure combinations in the thirty-six tests. The reliability of the complete system (engine and chassis) is that, on average, twenty-five out of thirty-six will pass, i.e. a reliability of 25/36 (or 5/6 × 5/6) = 0.694 or 69.4%.

Thus, even with only two parts the reliability of a system will be less than the reliability of any one part and, as has been seen, is determined by multiplying the two parts failure rates together, *not* by adding them.

Confidence levels

It will be remembered that reliability is defined as the *probability* that an item will perform a required function under stated conditions for a stated time. What confidence can be given to this probability? It can vary between 0 (none) and 100 (certainty).

If an item is for general use, the confidence level (CL) may be reasonably low (say 60%), but for equipment involving life safety may be much higher (say 99%).

The degree of confidence is tied up with statistics, and this, in turn, is concerned with the amount of evidence—in particular, the size of the sample taken for testing. The effect of confidence levels on reliability calculations can be great. For instance, for an electronic equipment to have a failure rate of 1% per 1,000 hours, the minimum test hours required with no failures for 60% confidence level would be 92,000, but for 90% CL, 230,000 unit test hours would be required.

Sampling schemes are also used for tests in which a certain number of a total quantity (or population) are selected according to the quantity and the acceptable quality level AQL determined, but these are again discussed in references 1, 3 and 8 (see page 59).

The language of reliability statisticians

Exponential, Normal or Gaussian, Poisson, Gamma and Weibull distributions are commonly used in estimating reliability. The reader will, however, come across a number of unfamiliar terms which are used by statisticians, such as contingency, goodness of fit, Chi-square (χ^2) distributions, "students t" variance ratio, etc. Some of the language used by modern statisticians, is defined below for interest:

(1) *Algorithm*
A prescribed set of well-defined rules or processes for the solution of a problem in a finite number of steps.

(2) *Cut sets and Tie sets*
Cut set is a set of components in a system whose failure results in the system failure.
Tie set is a set of components in a system whose success ensures system success.

(3) *Duane Model*
A model which provides deterministic approach to reliability growth such that the system MTBF versus operating hours falls along a straight line when plotted on log–log paper.

(4) *Fault Tree*
A fault tree is a graphical representation of logic associated with the development of an undesired system event in terms of the factors that contribute to its occurrence.

(5) *Fuzzy sets*
Fuzzy sets refer to a class of set with continuous grade of membership involving a gradual (rather than abrupt, as in case of conventional sets) transition from membership to non-membership.

(6) *k-out-of-m:G*

The system consists of *n* redundant units of which at least *k* units should be good for the system to be good.

(7) *k-out-of-m:F*

The system consists of *n* redundant units of which at least *k* units will fail before the system will fail.

(8) *Heuristics*

These pertain to exploratory methods of problem solving in which solutions are devised by evaluation of the progress made towards the final result. These are generally discovered in-tuitively.

(9) *Markov-Chain*

When the behaviour of a system is described by saying it is in a certain state at a specified time, the probability law of its future states of existence depends only upon the present state and not on how the system reached that state, the system state behaviour can be described by a process called the Markov process. A Markov process whose state space is discrete is called a Markov chain.

(10) *Modelling*

A technique of system analysis and design using mathematical or physical idealizations of all or a portion of the system based on the state of knowledge of the system and its environment.

(11) *Petri Nets*

A bipartite directed graph modelling containing places, transi-tions and tokens used for studying structure and control of a complex concurrent information processing system.

(12) *Stochastic analysis*

An analysis or procedure employed for the determination of certain parameters of stochastic/random process, using some system modelling equations and boundary conditions.

(13) *Truth Tables*

A table that describes a logic function by listing all possible combinations of states of system component and indicating, for each combination, the state of system.

CHAPTER 4

How Reliability is Calculated

It is important to remember that one cannot calculate the exact period for which an equipment will work without failure. All that can be done is to calculate the *probability* of an equipment working without failure for a particular period of time.

The concept of probability is fundamental to an understanding of what is meant by reliability, so it is vital to be quite clear just what "probability" means.

Suppose we have a bag containing an equal number of well-mixed black and white marbles. If you put your hand in without looking and take out the marbles one by one, the probability would be that if you took out a certain number you would remove equal quantities of black and white marbles. The more marbles you took out, the higher the probability that the quantities of each would be equal. But this is only a probability; it is what we expect to happen. But we cannot rule out that you might, for instance, take out all the black marbles first.

Again, if the bag contains twice as many white marbles as black, then the probability would be that after taking out a certain number of marbles you would have removed twice as many white marbles as black. Here again, the more marbles you took out the higher the probability that you would have removed twice as many white as black, but there is no certainty that this would be so.

In reliability, the average period an item will function without failure is most often represented by the mean time between failures (MTBF). But there is no certainty that it will not fail before the end of this period; it might fail within a much shorter period, or perhaps run for longer without failure. But as with our marbles, the longer we go on, the more likely is our probability to prove true taken over the whole period; that is to say, if an item is run for a period which is long compared with its MTBF, the more likely is the MTBF to represent the true average period between failures. But note carefully that this is not the same thing as saying that the period between individual failures will approximate to the MTBF more closely at the end of a long run than it did at the beginning.

The MTBF has, however, become a useful measure of reliability for comparison purposes. There are two ways of calculating or predicting this measure of reliability. Taking, as an example, electronic equipments:

(i) By estimating failure rates for all the parts in the equipment, multiplying these by the number of parts, then relating this to the total equipment MTBF.

(ii) From previous experience with similar equipment, set up equations which can be used to predict an MTBF for a new equipment.

The first system is the one most used and the conventional method for expressing the failure rate of parts is in percentage per thousand hours (%/1,000 hours).

Taking a detailed example, let us list the parts in a typical transistorized equipment before integrated circuits were introduced and their possible failure rates as shown in Table 4.1.

TABLE 4.1

Part	Failure rate (f) % per 1,000 hours	No. in use (n)	Product of (n) × (f)
Transistors	0.003	100	0.3
Diodes	0.002	200	0.4
Connectors	0.15	5	0.75
Resistors	0.001	150	0.15
Capacitors	0.006	100	0.6
Switch	0.03	1	0.03
Relay	0.04	1	0.04
Sockets	0.03	6	0.18
Potentiometers	0.06	20	1.20
Transformers	0.04	2	0.08
Total		585	3.73

From Table 4.1 the total equipment failure rate is thus 3.73% per 1,000 hours of operation. The relation between failure rate and MTBF is

$$\text{failure rate} = \frac{1}{\text{MTBF}}$$

$$\text{MTBF} = \frac{1}{\text{failure rate}}$$

So in 1,000 hours 3.73 out of 100 parts may be expected to fail. So

$$\text{MTBF} = \frac{1}{\text{failure rate}}$$

$$= \frac{1}{3.73 \times \dfrac{1}{100,000}}$$

$$= \frac{100,000}{3.73}$$

$$= 26,810 \text{ hours.}$$

This is the simplest form of prediction of MTBFs and in practice many other factors have to be considered—the component part stresses, the environmental conditions, special operating conditions, etc., but these are usually taken into account in the design. The total number of parts used has a bearing on reliability, as obviously an equipment with 2,000

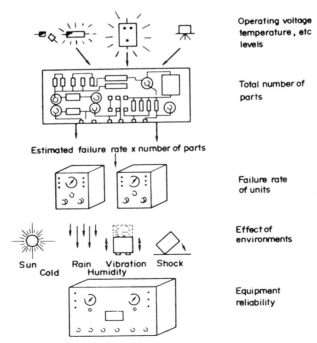

Operating voltage temperature, etc levels

Total number of parts

Estimated failure rate x number of parts

Failure rate of units

Effect of environments

Sun   Cold   Rain   Humidity   Vibration   Shock

Equipment reliability

FIG. 4.1. Factors taken into account in estimating electronics reliability.

parts will have a higher failure rate than one with 200 similar parts assuming they are operated under comparable conditions.

The main factors taken into account in calculating reliability are illustrated pictorially in Fig. 4.1.

Typical failure rates for electronic parts are discussed later in this chapter, but it must be emphasized that these are given as a guide only. They will vary with component stresses, environment, operating conditions, etc., and will change with any alteration in materials or processing employed in the manufacture of the component, hence the necessity of continually up-dating failure rates.

Some values of MTBF are given on page 22.

The number of component parts used varies widely. For a television set 500 to 1,000 parts may be used, for a large computer 250,000 to 500,000, while for space operations such as a manned moon mission up to 15 million parts may be required, most of these are now combined in the form of integrated circuits, or "chips".

Physics of failure

A difficulty in estimating reliability statistically is the very large number of components which must be tested to get a meaningful number of failures. For instance, if 330,000 specimens of a component with a failure rate of 0.001% per 1,000 hours were tested for 1,000 hours (nearly 42 days) there would be only three failures on average. It is unlikely that the production run would be large enough to justify testing on this scale, and the time, equipment, and expense involved is considerable.

For high reliability items such as integrated circuits a new approach has therefore been developed, termed the physics of failure, which sets out to prevent the manufacture of faulty components, in contrast to detecting and rejecting them. The integrated circuit itself is extremely reliable; faults occur in encapsulating it, mounting it, and connecting it to lead-out wires. Through careful investigation and analysis, faults in these areas can be avoided.

Failure rates of electronic components and microelectronics

As a general guide, fixed resistors and fixed capacitors have fairly low failure rates of the order of 0.005–0.05% per thousand hours; transformers about 0.01–0.05% per thousand hours for each winding. Relays have to be calculated on failures in contact pairs and on coils and depend on many factors (as each type of relay has a different failure rate). Higher failure rates are expected from variable resistors, approximately 0.02–0.06% per thousand hours, again depending on type, whilst switches are rated on contacts, being approximately 0.015% per thousand hours per contact. Valves used to have much higher failure rates, ranging from 0.5% to 3.0% per thousand hours, as also have some mechanical devices such as blower motors (4% per thousand hours). Connections, although having extremely low failure rates, are numbered possibly in thousands so that hand-soldered connections may have a failure rate of 0.001% per thousand hours whilst machine-soldered connections can be as low as 0.0005% per thousand hours. Crimped connections are approximately 0.002% and welded 0.004% per thousand hours, whilst wrapped connections are regarded as the most reliable, having a very low failure rate, of the order of 0.0001% per thousand hours. Transistors range from about 0.003–0.01% per thousand hours.

Integrated circuits (ICs) or "chips" are now used in all electronic equipments.

The introduction of silicon integrated circuits in the late 1950s and early 1960s produced a revolution in electronics so far-reaching that it can be likened to the invention of printing from movable type in the late 15th century.

With the virtual elimination of the thermionic valve (vacuum tube) and of large numbers of individual resistors, capacitors, etc.—reliability, size, weight and cost were greatly reduced. This made possible the production of complex modern digital computers, pocket calculators, very complicated airborne electronic equipment and other innovations.

Because of the multiplicity of technologies employed in integrated circuits (TTL, ECL, MOS, MNOS, CMOS, etc.) and differing encapsulation techniques it is impossible to give any overall generic failure rates for integrated circuits. In the first 15 years after their introduction, however, the failure rates for all types of device decreased by approximately one hundred times. This reduction in failure rates is still continuing and reliability calculations should use only the most up-to-date figures available. For this reason no attempt is made in this book to tabulate precise failure rates as any values given would soon be outdated. One example that may offer some indication of present trends can, however, be quoted. A very complex integrated circuit containing over 14,000 "components" (some 800 gates) is reputed to have a failure rate of 0.0339% per 1,000 hours at a junction temperature of 125°C. Less complex circuits or circuits operated at lower junction temperatures would have a considerably smaller failure rate.

To summarize—reliability assessment procedure

To calculate MTBFs list the component parts accurately, find out the basic failure for each, and multiply this by the number of components. Add all products and divide into 100,000. This estimates the MTBF of equipment in hours. The basic failure rates are modified by a weighting factor (due to the environment in which the equipment is operated) and a possible rating factor (depending on whether the components are fully loaded or not), and also temperature. It is quite difficult to give an accurate estimate but it is often necessary to predict an approximate MTBF in the early stages of equipment design. For the sake of simplicity and uniformity, failure rates have been given in % per 1,000 hours (i.e. failures in 10^5 hours) in this book, however failure rates may also be given as failures in 10^6, 10^8, 10^9 or 10^{12} hours.

Derating to improve reliability

Manufacturers' ratings are usually based upon safe working loads for a reasonable performance. A bridge may be rated to carry 30 tonnes, but could probably carry 60 tonnes. A mechanical hoist may be rated at 10 tonnes, but could be used for more than this. A 5-watt resistor could be loaded for more than 5 watts, but would not be so reliable. A balance must always be obtained between rating, cost and reliability.

Values of MTBF

Some approximate MTBFs for electronic equipments are given in Table 4.2 as a rough guide, but again conditions vary considerably and are different for each individual equipment.

TABLE 4.2

Approximate MTBFs for electronic equipments

Equipment	MTBF (approximate)
Repeater amplifiers in undersea cables	40 years
Satellites (some)	10/20 years
Computers and electronic equipment in laboratories	5,000–10,000 hours
Military ground equipment	1,000–5,000 hours
Shipboard electronics	500–2,500 hours
Airborne electronics	100–1,000 hours
Missile electronics	1–500 hours

MTBFs for lifts, radios, TV sets, refrigerators, washing machines, cars are not readily available, but an idea of approximate MTBFs is as follows:

Lifts (elevators)   approx. 5 years (or 44,000 hours).
TV sets   approx. 10,000 hours.

Refrigerators   approx. 30,000 hours.
Washing machines   approx. 10,000 hours.

Life expectancy = mean life ≠ MTTF

It is of interest to compare the life expectancies of some well-known objects with that of a modern integrated circuit.

Butterfly   2 weeks
Light bulb   1 year
Motor car   10 years
Undersea cable   40 years
Man   75 years
House   100 years
Single integrated circuit   Hundreds of years

Redundancy

In general, redundancy means more than one way of accomplishing a required function. It can take many forms: partial, active, standby, etc.

Partial redundancy can be illustrated by spokes in a bicycle wheel; even if some of them are broken the wheel will not fail. A lift suspension may consist of more than one rope so that in case one fails the lift can still operate. A further example is that of a four-engined aeroplane, capable of taking off on three engines.

In the case of equipment used in space missions, satellite communications, complex military systems, etc., when failure would have far-reaching effects, defence, safety or financial consequences, redundancy is employed. This means that part or the whole of a particularly vulnerable equipment is replicated in order to reduce the probability of loss of system function. A simple example of redundancy is the use of stand-by electrical generators in a factory or hospital to ensure continuity of supply in the event of a mains failure. A system of this type is known as stand-by redundancy, i.e. in the event of the failure of A, then B, a replacement device, can be switched into the system either manually or automatically (see Fig. 4.2).

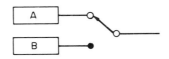

FIG. 4.2. Stand-by redundancy.

Another form frequently employed particularly in signal processing applications is "parallel redundancy". In Fig. 4.3, A and B are identical devices, both capable of performing their function independently. In the event of failure of either of the elements A or B, the overall system function will not be lost.

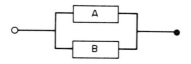

FIG. 4.3. Parallel redundancy.

Many other configurations are possible; a parallel system of say three or more elements where the overall function is unaffected by the loss of one element, but the loss of two elements may result in reduction of function but not complete loss.

It is obvious that in the employment of redundancy techniques and therefore duplication of system elements both the size and cost of an installation are increased. This limits its application to cases of absolute necessity. One typical example of a situation where redundancy is highly desirable, but in practice is frequently prevented by space/weight considerations, is that of the transmitter in an airborne radar installation. The transmitter is usually the area of greatest reliability hazard in any such system because of the particularly severe environment associated with high power (hence high temperature) and high voltage combined with vibration. The limited space and load carrying capacity available in most aircraft prohibit the replication of transmitters in the majority of cases.

The calculation of MTBF for a system employing redundancy is obviously complex and is beyond the scope of this book; however, guidance on the mathematical techniques required can be found in most standard works on reliability.

Reliability in manufacture

By the time an equipment has passed from development into the production stage, most, if not all of the design deficiencies should have been eliminated. Under ideal conditions only random component failures would then be experienced and the MTBF should be that produced by the prediction calculation; this represents the full reliability potential of the equipment. Any errors in manufacture, e.g. badly made soldered joints, locking devices or materials, screws omitted, etc., will reduce the MTBF from the predicted value. "Burn-in" can, and does eliminate many of these problems, but in practice it has been found that a considerable number of manufacturing errors required a considerably longer time than that economically possible for burn-in, before they cause failure. For this reason, strict quality control including rigorous inspection is essential if the full reliability potential is to be achieved.

Quality control

The quality of any manufactured product is determined by its design, the materials from which it is made and the processes used in its manufacture. The BSI formal definition of quality is "The totality of features and characteristics of a product or service that bear on its ability to satisfy a given need". This means that the item must meet its specification when it is required by the user. How well it meets its specification is called "conformity", defined as "The ability of an item to meet its stated performance and/or characteristic requirements, the assessment of which does not depend essentially on the passage of time". This is the "zero" time element of quality, whereas the "future" time element is the reliability.

Acceptance testing is mainly a question of economics. To prove a high degree of conformity requires a great deal of testing and therefore is very costly, and this must be balanced against the value of trouble-free operation by the customer. As mentioned in Chapter 1, this importance varies widely and the cost of quality control must be carefully estimated.

Acceptance testing is carried out by taking a sample from every batch manufactured and subjecting it to a wide range of measurements under various conditions. The size of the sample selected varies according to the AQL value chosen. The AQL (acceptable quality level) is the maximum average percentage of defective items which is considered tolerable. The lower the AQL value, the larger the sample must be. Tables are provided known as "sampling tables" which statistically give the number to be tested, and the number accepted (which pass all the tests), against the AQL value.

Quality of a product, therefore, is determined by conformity and reliability.

Calculation is not everything

This chapter has dealt with calculating the reliability of components, and has touched on the reliability of equipments and of systems. However, for some designs there is an additional factor which the engineer may need to take into account in assessing reliability and safety. This additional factor is public opinion, a variable, emotional, non-quantifiable factor, set smouldering by major accidents to nuclear installations, aircraft and ships, and often fanned into flame by the media. Public opinion can overwhelm calculated engineering design, as was shown in 1988 by the closure of a nuclear power station on Long Island in the U.S. before it had ever generated even one watt of electricity. The closure was forced by public opinion on the major grounds that if there were an accident in the installation it would be impossible to vacate Long Island quickly enough.

Stages in producing a typical electronic equipment

Before discussing the effect of operating conditions and environments on electronic equipments in the next chapter, it must be emphasized that the design, development and production of any electronic or mechanical equipment is extremely complex.

To illustrate this, the following chart has been produced which not only shows the various stages in design, development and production, but also the stages necessary in ancillary and reliability work. Even with computer designed equipment, with assembly by robots, similar complex work has to be carried out from the initial idea to the final filing of documents.

Ancillary work required (1)	Stages in design, development and production (2)	Reliability work required (3)
	Initial idea	
Market research	Tentative requirement	
Initial calculations	Design proposals	
Literature review	Initial specification	Estimate likely reliability requirement
Initial component part choice	Circuit design	Determine likely environments / Use "worst-case" design / Select likely best component parts / Investigate redundancy
Tests and modifications	Breadboard model	Estimate approx. MTBF
Component part lists prepared	Design approval	
Development component parts ordered	Development model	Evaluate component part reliability / Establish possible failure modes / Estimate MTBF
Drawing office work started	Mechanical design	Mechanical tests—Resonance search / —Bump tests
	Layout of units and system design	Best ergonomic layout / Adequate cooling / Good accessibility for maintenance / Best interconnection

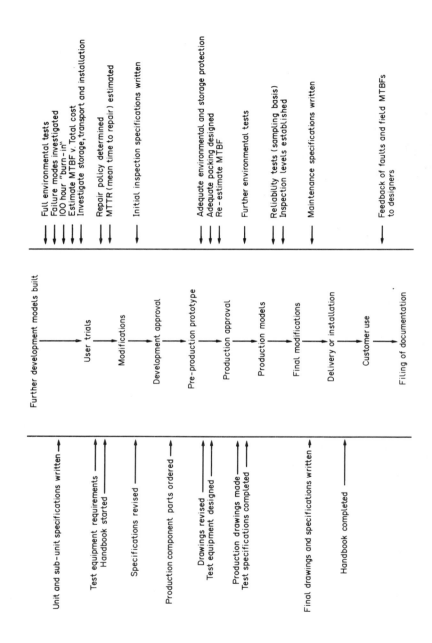

CHAPTER 5

The Effect of Operating Conditions and Environments

Equipments have to be installed and operated in all parts of the world, and climatic conditions can have a considerable effect on reliability. For instance, in hot, wet climates such as the Tropics the humidity is very high. In desert areas the temperature is high and the humidity low. In the Arctic the temperature is low. Some of the effects of these and other environments on the number of failures of equipments and materials is shown in Table 5.1.

TABLE 5.1
Analysis of Typical Failures of Military Equipment
and Materials under Various Environments in a Given Time

Environment	Temp. and humidity	Dust	Humidity	Radia-tion	Salt spray	High temp.	Low temp.
Electronic and electrical equipment	17	3	10	1	–	14	14
Lubricants, fuels and hydraulic fluids	–	1	–	–	–	8	4
Metals	10	–	9	–	26	8	5
Optical instru-ments and photo equipments	5	1	3	–	–	–	–
Packaging and storage	9	–	9	–	–	–	–
Textiles and cordage	14	–	5	11	–	3	–
Wood and paper	12	6	4	–	–	2	–
Totals	67	11	40	12	26	35	23

It will be seen that humidity heads the list of factors causing failure. This is because moisture can form films on insulating materials, producing electrical losses, penetrate into component parts and change their characteristics and also cause rust to develop on a variety of metals, reducing their mechanical strength. Equipment for use in the Tropics is therefore usually sealed up (although sometimes this produces heating problems) or heated continuously to drive off the moisture.

Operating in hot climates also produces problems. Most electronic equipments dissipate some heat and if to this is added the climatic temperature the total heat may affect the reliability of the equipment. Greases may melt and some low melting point plastics such as polythene may soften. Differential expansion of differing materials can cause distortion and binding of moving parts.

Low temperatures also have an effect on electronic and other equipments although in most cases equipment is operated in heated rooms because of the human operator. There are, however, some instances where it is not possible to heat the equipment, such as walkie-talkie sets and equipment in aircraft.

The effect of these environments on the failure rates of electronic equipments may be summarized and are shown in Table 5.2.

TABLE 5.2
The Effect of Environment on Electronic Equipment Reliability

Equipment	Environment	Approx. failure rate (% per 1,000 hours)
Underseas cable amplifiers	Extremely stable temperature, low humidity, no shock or vibration once laid	0.0004
Electronic telephone exchanges	Usually air-conditioned and well maintained	0.067
General-purpose computers	At room temperature Some air-conditioned	0.17
Civil airlines	Subject to vibration and temperature changes but well maintained	0.83
Laboratory equipments	At room temperature subject to handling	1.01
Radio and TV sets	Room temperatures. Not usually moved when installed but built at competitive prices	1.00
Military equipment	High temperatures High humidity Salt laden atmosphere Shock, vibration, etc.	2–4

Burn-in

Reference back to Fig. 2.1, page 7, shows that during the early life of components, the failure rate is considerably higher than during the useful life period. If components and sub-assemblies were incorporated directly from the production line into a complex equipment which was then shipped to the customer, an unacceptably high failure rate might be experienced during the first few hundred hours. In order to minimize this effect, a procedure known as "burn-in" is frequently employed. The individual components (integrated circuits, transistors, diodes, capacitors, etc.) or sub-assemblies, e.g. printed circuit boards are operated for a hundred or so hours under environmental conditions slightly more severe than those encountered during normal operation. The effect of this is to induce many of the "early life failures" in the components or sub-assemblies before system integration. This is obviously a fairly costly operation, but a large overall saving is possible due to the reduction of the requirement for re-work on the final product and also customer goodwill is maintained.

Burn-in of the completed equipment is possible, but the diagnostic and re-work problems encountered can lead to a considerably larger expenditure in time and money than if the operation is conducted at a component level.

CHAPTER 6

Mechanical Reliability

The well-reported failures, such as the Space Shuttle Challenger, the Three Mile Island and Chernobyl nuclear accidents, and the Bhopal gas escape, emphasize vividly the necessity for mechanical reliability.

Buildings, bridges, transit systems, railways, automotive systems, robots, off-shore structures, oil pipe lines and tanks, steam turbine plates, roller bearings, etc., all have their particular modes of failure affecting their reliability.

Whilst this book deals mainly with electronics, there are a number of common modes of mechanical failures, which are worth listing, e.g. with structures:

(1) Corrosion failures
(2) Fatigue failures
(3) Wear failures
(4) Fretting failures
(5) Creep failures
(6) Impact failures

These may be considered the main failure modes, but there are of course many others, such as ductile rupture, thermal shock, galling, brinelling, spalling, radiation damage, etc., which are outside the scope of this book.

General mechanical failures

There are many causes of failure. Some have been listed by C. Lipson* as:

(1) Defective design
(2) Wrong application
(3) Manufacturing defect

---

* Lipson, C. "Analysis and Prevention of Mechanical Failures," Course Notes 8007, University of Michigan, Ann Arbor, June 1980.

(4) Wear-out
(5) Incorrect installation
(6) Failure of other parts
(7) Gradual deterioration

Some interesting sources of product failure in the following areas are given by Lipson* as:

(1) Steel industry. Manufacturing and metallurgy (34%), design (58%), and service (8%).
(2) Engines. Manufacturing (20%), engineering (40%), misuse in field (30%), and other causes (10%).
(3) Truck industry. Manufacturing (12%), design (55%), and materials (33%).
(4) Electrical industry. Manufacturing (37%), design (37%), and service (26%).

Guidelines for designers in order to achieve higher reliability are given in B. S. Dhillon's book: *Mechanical Reliability: Theory, Models and Applications.*†

(1) Make design as simple as possible.
(2) Avoid introducing cost-saving procedures at the expense of reliability.
(3) Make allowances for human error.
(4) Make use of well-tried parts and materials.
(5) Carefully test new designs.
(6) Perform analyses of data received from the field.
(7) Pay attention to data received from the field when making modifications and improvements.
(8) In the case of critical parameters, consider using safety margins of 3–6 standard deviations.
(9) Consider diagnosis of critical items.
(10) Include facilities for inspection in the design.
(11) Introduce redundancy whenever necessary.
(12) Pay attention to the maintenance aspect with respect to reliability.
(13) Take into account the effects of transport, handling, and storage.
(14) Make use of standard parts as much as possible.

---

* Lipson, C. "Analysis and Prevention of Mechanical Failures," Course Notes 8007, University of Michigan, Ann Arbor, June 1980.

† B. S. Dhillon. *Mechanical Reliability: Theory, Models and Applications.* AIAA Education Series. 370, L'Enfant Promenade, SW, Washington, DC 20024, United States of America. Published 1988.

(15) Pay attention to the manufacturing aspects with respect to reliability.

Some of these are touched upon in sections of this book, e.g. Chapters 5, 6, 7 and 8.

Human reliability is often discussed and it is a fact that, apart from war, the highest toll of human life is caused by car accidents. These, in turn, are mainly caused by errors of judgement, making the main cause of failure—lack of human reliability

CHAPTER 7

Installation and Operability

Installation

The ways in which the installation of an equipment or machine can affect its reliability may not be immediately apparent, especially since the method and means of fixing and putting into operation will have been largely laid down beforehand. There may be some latitude in the running, connecting up, and inter-connecting of the necessary services such as electricity, air, water, etc., so that the following two points, for instance, must be watched.

(1) Care must be taken to ensure that pipes, cables and flexible connections will not be damaged by any vibration which may be present. They can be damaged by chafing against a support, against the machine, or against each other. For instance, electrical connections passing through access holes or supports must be protected by grommets or other means.

(2) Cables which run across the floor must be positioned so that they are not trodden on, either during installation or subsequently. This is especially liable to happen if the equipment or machine is being installed in a confined space.

If installation has to be carried out under adverse conditions the possibility of errors increases. In planning how an equipment will be installed, full attention must, therefore, be given to the environment in which the work will be done. Since installation involves the same type of operations as are performed in repair and maintenance, similar considerations of human engineering and of working conditions are involved. These are described in this chapter.

Operability

Neither operability nor maintainability (dealt with in Chapter 8) are features which are usually considered within the scope of reliability.

Nevertheless, poor operability can lead to equipment failure, and poor maintainability can reduce the time during which an equipment is functioning normally. For these reasons both operability and maintainability have been included in this book.

Every equipment has to be operated by a human operator. If the equipment does not perform its specified function, the result is a failure or lack of reliability. To be reliable equipments must, therefore, be so designed that the probability of the operator making an error is as small as possible, and this characteristic is called Operability.

To produce a design best suited to a human operator we must know the relevant facts about human capabilities and limitations. These facts come from investigations in the field of "ergonomics", i.e. the study of man in relation to work. The design and specification of a human being, that is to say the way in which he controls and moves his limbs, reacts to various stimuli, and so on, are fixed, although human beings vary considerably from one to another and are also very adaptable. This specification must be carefully considered in designing a machine or equipment for human control, and ergonomics or "human engineering", as it is sometimes called, lays down the rules which enable the best possible use to be made of the working human being.

For instance, standing up is tiring, so the operator should be seated whenever possible. When seated the height and design of the seat, backrest, footrest and armrest, the height of the working area and the position of the various parts which the hands and feet must reach must all be considered in relation to the job to be done and the average dimensions of the people doing it, in order that fatigue should be reduced to a minimum. As an example, the optimum distance to place controls is about 70 cm from the operator.

In cases where physical effort is required to operate controls, ergonomics has established rules which enable a man to supply the effort with a minimum of fatigue. These rules lay down, for instance, that it is uneconomical to underload large muscles and dangerous to overload small ones; that however perfect a machine may be, it is not correctly designed if only an athlete or contortionist can operate it; and that needless fatigue must always be avoided. Some of these rules may seem obvious, but are nevertheless often overlooked.

Ergonomics also provides guide lines on the most comfortable working environment. For instance, a high noise level is extremely tiring, and high-pitched sounds are much more disturbing than low-pitched sounds. The maximum tolerable noise level is around 90 dB; at this level it is difficult to talk to a person 1 metre away. Two simple rules apply here—put the noisy parts of a machine as far from the operator's ears as possible, and find out when ordering an equipment what is its maximum specified noise level. Noise can be reduced by replacing worn components, by damping out vibration, by insulating the source of noise, or by insulating noise-reflecting surfaces such as walls and ceiling.

Another important consideration in the working environment is lighting. Either too much or too little can be tiring. Too much lighting can be caused by glare from highly polished or other reflecting surfaces, or from bright lights within the field of vision. Bright lights can be particularly tiring; windows fall into this category, and positioning an operator directly facing a window when viewing a control panel should be avoided. The need for adequate lighting is obvious, but the actual level and nature of the source must be related to the position and type of controls and indicators with which the operator is concerned, particularly those who spend long hours at video display units (VDUs) and plan position indicators (PPIs).

The human body is fairly sensitive to its environment. Muscular effort will generate heat; if the operator exercises little effort he may need heated surroundings. Body temperature is controlled through convection (the transfer of heat to or from the surrounding air) and through radiation (the transfer of heat to or from surrounding surfaces such as walls and windows). To make the operator comfortable both these conditions must be controlled. Attention must also be given to air velocity which may result from draughts, or from wind in outside locations, and to humidity. Humidity is related to body temperature, since excessive humidity reduces the body's ability to lose heat through the evaporation of sweat and so reduces its resistance to high temperatures.

Another source of human fatigue is vibration, which may be encountered particularly in large machines, or in aircraft, ships, or large vehicles. Vibration is tiring because it causes relative movement between the body and the points of contact with the surroundings, or between different parts of the body itself. Thus the effects of vibration depend on the nature of the vibration, and on its frequency. For instance, vibration may be periodic, as in aircraft, or non-periodic as in a ship in a rough sea. Intense and continuous vibration between 0.1 and 1 cycle per second can produce a form of sea sickness. From 1 to 15 or even 30 c/s is the range of greatest importance, because it is here that the chief resonances occur in the human body; it is also the band of frequencies in which many machines and vehicles vibrate, and in which vibration is particularly difficult to suppress. Vibration above 30 c/s is generally not a serious hazard unless particularly intense; its effects are somewhat similar to noise.

To avoid excessive fatigue due to vibration the nature of the vibration must be predicted or measured, and an acceptable level defined. Both these steps may involve complex matters of design engineering, and the second may involve the intricacies of human engineering. The final step is to control the vibration in order to bring it into acceptable limits, and this again may be difficult and require considerable expense.

To sum up, the more comfortable the operator is, the less likely are mistakes which will introduce an element of unreliability into the equipment being controlled because, through faulty handling, it may fail to perform its specified function.

Ensuring correct operation, however, involves much more than just making the operator comfortable. The operator's main responsibility is to assess an equipment's performance through displays such as meters, position indicators, counters, or lamps, or in the case of vehicle control, for instance, through the assessment of external conditions, and control it through the movement of knobs, levers, etc. The displays must be so designed that the operator will quickly and accurately understand the information they convey, and the controls must operate in a way which will as far as possible ensure that when the operator moves them, the required change in the performance of the equipment will result.

The operator and the equipment in fact form a closed loop in which information output on the machine is received by the operator's input (eye or ear). The operator's output is in the form of an action which feeds into the machine input and operates a control mechanism which alters the performance of the machine in a manner which is again shown on the information output or by a change in its performance. Any failure in this loop is likely to result in the equipment failing to function correctly, thus introducing an element of unreliability. The way in which the equipment responds to its control mechanism is a function of the equipment design, and so is not relevant to operability. The overall reliability of an equipment thus depends on the integration of the equipment and the operator, and this means considering the type and position of information displays, and the position, movement, and type of controls in relation to operator comfort and movement.

Mistakes in operation will also be reduced if controls are grouped according to their function, and according to the particular part of the equipment which they influence. Controls which have to be operated in sequence must be placed near each other, and in the correct order. The various requirements to reduce operator errors can never be met by a symmetrical layout of controls with identical knobs. Unfortunately this type of panel, which is designed to look right and not designed for good operability, is still the rule rather than the exception.

Having received information from the display, the operator must move a control. It is essential that he moves the correct control in the right direction, and the following are the more important considerations which will increase the probability that this will be done.

A similar layout for display and controls will assist in selecting the correct control. A further advantage is for the operator to associate a particular shape of knob or control with a particular function, by touch as well as by sight.

When a control is moved it is essential that the operator can see the associated display. It must not, for instance, be placed round a corner or behind the operator's back.

The position in which a control is set must be clearly and quickly distinguishable. If the positions of a rotating control are numbered, then laying out the numbers in the same positions as on a clockface enables the operator to tell instinctively what position the control is in.

Complex systems contain many visible displays and some audible signals to indicate and warn of incorrect operation and of fault conditions in the equipment. In the event of a serious fault there can be so many warning signals that the operator can be presented with much more information than can be assimilated and acted on correctly under emergency conditions; for instance, in a nuclear power station. The operator is consequently unable to distinguish the actual fault and the malfunctions arising from the fault, from the mass of warning information being presented. Designing a warning system which will be effective in all fault conditions is therefore complex and difficult, and must be closely related to human characteristics; it will involve such considerations as the nature and positioning of warning signals, the priority in which they require attention under various fault conditions, and the ability and speed with which an operator can be expected to respond to them correctly in an emergency.

The type of control chosen must be related to its function. Levers should preferably be long, so that less force is needed to move them, and the wrist and forearm should be supported if fine adjustment is required. Cranks are most suitable where a wide range of adjustment is needed, because they can be turned quicker than a wheel or a knob. Handwheels should only be used if considerable force must be transmitted.

To sum up. The critical moment in the operation of an equipment is when some decision is taken, and as a result some control is moved. The reliability of the equipment is bound up with the decision the operator makes, the actual control he operates, and the direction in which he moves it, and these can all be affected by his personal comfort and working conditions, in addition to his interpretation of a display, and the layout and design of controls.

CHAPTER 8

Maintainability

When an equipment or machine fails, as it inevitably will sooner or later, since no design can be made absolutely reliable, it is important that it should be repaired quickly so as to become available for use again in the shortest possible time. Often, in fact, the user is more interested in availability, which is the percentage of time during which the equipment is functioning when required, as in television sets in the home.

Equipments must, therefore, have good maintainability, maintainability being a measure of the speed with which loss of performance is detected, the fault located, repairs completed, and a check made that the equipment is functioning normally again. Maintainability must be built into the original design since, like other reliability factors, attempts to incorporate it as an afterthought, by modifications to the manufactured equipment, will never produce a satisfactory solution. Features which improve maintainability can often be made part of the original design without a great deal of extra expense. The designer must, however, be careful not to improve maintainability by introducing features which reduce reliability—for instance, by enabling assemblies to be removed quickly by replacing soldered connections by plugs and sockets which are sometimes less reliable.

To obtain satisfactory maintainability the following factors must be considered:

(a) The equipment or machine will fail at some time or other.
(b) The positioning of maintenance check points, gauges, meters and the position of one assembly with respect to others.
(c) The limitations imposed by the human frame.
(d) The environment in which maintenance or repairs will be carried out.
(e) The design of test equipment.
(f) The presentation of information in the maintenance and repair manual.

Good maintainability is just as important for routine maintenance as it is for repair, since maintenance can also represent a period of non-availability which must be made as short as possible.

Maintenance involves making a number of check measurements, or taking specified action such as lubrication, at particular points in the equipment or machine. It is essential that these points are easily accessible and are so situated that the required action can be easily taken. The check points must not be put in positions dictated by the physical design of the equipment rather than in the best position for maintenance. If they are, they may be difficult to get at, or be so placed that when an adjustment is made or a reading taken it is difficult or impossible to see the relevant meter or indicator. It is essential that all the displays, meters, component parts, adjustment points or other items which will be involved in a single maintenance operation should be so grouped that all are comfortably accessible, both manually and where the eyes can easily be placed in line with the pointer, to ensure an accurate reading.

Special attention must be given to the positioning of parts which must be regularly serviced, or which are known to have a high failure rate. They should be placed so that the maintenance points and the parts themselves are readily accessible, without first removing other components or assemblies. Covers on the part itself, or which must be opened or removed to get at the part, must be easily and quickly removable. Obviously replacement is quicker if a part can be replaced from the front of an equipment—a pilot lamp, for instance. Where there may be difficulty in identifying a part such as in electronic equipment, the part must be clearly marked in a place which is easily visible.

If colour coding is used, it must be remembered that about one in seven persons are sufficiently colour-blind to be unable to distinguish nearly identical shades, or dull colours. Colours must not fade, they must be used consistently, and whenever possible a standard code should be used, details of the code being displayed on the equipment. The eye finds it difficult to identify small areas of colour, so that the use of, for instance, a colour thread running through cable insulation is unsatisfactory. The use of colour coding on parts which are liable to get dirty or contaminated with oil should also be avoided.

Fixing devices must be readily accessible, so that a part can be removed easily and quickly for repair or replacement. Fixing devices must also be designed so that a part can be quickly released and quickly re-secured.

It must be remembered that what appears completely accessible, or very simple to adjust, or to undo and do up, when a part or assembly is on a bench may not turn out to be so when assembled into a complete equipment. It is, therefore, essential to visualize, at the design stage, exactly how each part will be situated in the complete assembly, equipment or system. The designer must constantly aim to cause the minimum of work for the maintenance engineer. This involves not only making it easy to "get inside the black box" and doing the job when there, but also avoiding the use of non-standard or too wide a variety of fixing screws. Even such an apparently unimportant feature as the use of needlessly long screws, which increase the time taken to remove and replace the nuts, lengthens maintenance time unnecessarily and so reduces equipment availability.

Clearly, maintenance cannot be carried out so well, or perhaps cannot be carried out at all, if there are any actual or potential hazards such as high voltages or moving parts. If interlocks have to be made inoperative during maintenance or repair, a warning light should indicate if power is on. In the case of machinery, some means should be provided to prevent inadvertent operation.

Good maintainability also involves consideration of the conditions in which the human body can work best, such as the relation between manual ability and the senses, human lifting capacity, and the maintenance equipment environment. Our work skill actions are governed by the input of information received from our senses, the most efficient input coming from sight, then touch and lastly memory. This means that work is performed most quickly and accurately when one can see what one is doing; if it is only possible to work by feel the job is more difficult, and if parts can be neither seen nor felt but their position has to be memorized, a job becomes lengthy and tedious. To take a simple example—it is easy to place the blade of a screwdriver into a slot on a screw head which can be seen, rather more difficult if the screw cannot be seen and the blade must be guided by hand, but very much more difficult when the hand holding the screwdriver must place the blade in the screw slot aided only by what can be remembered of the screw and slot position.

It is a matter of personal experience that in certain positions of the body the hands can carry out certain movements easily, either when holding tools or by themselves. Equipment and machines should be so designed that maintenance and repair can be carried out with natural movements; unnatural movements of the hands, arms or body lead to strain, frustration, wasted time, and a greater possibility of error. In this connection it is especially important to have in mind how a part or assembly will be situated within a complete equipment.

The maximum weight which a human being can lift varies within wide limits, depending on the actual position of the body and its position in relation to the weight. For instance, most men can lift 50 kg or more if it is close to their body, but only about one-quarter of this weight can safely be handled if it is positioned over 60 cm away. Less weight can be handled if the body has to work in a space which is restricted because access is difficult, or because the equipment is installed in a small area. Lifting capacity will also be reduced in other circumstances—for instance, on a ship which is rolling.

These weight limitations must be considered in the equipment design, particularly in mechanical units, so that they are not exceeded by any single complete item which will have to be manhandled in or out of the equipment. Even when the weight is within the limits all but the smallest and lightest units must be provided with some convenient handle or other facility for lifting. Not only does this make lifting easier, but if no provision is made the assembly may be moved by grasping some portion of it not designed to withstand its weight, with consequent damage, or it may be dropped.

Equipment is normally designed, built, and tested in an environment with a comfortable temperature and good lighting. However, maintenance and repair may have to be carried out in a hostile environment. For instance, in the case of vehicles, ships, aircraft and equipment used by the Armed Forces, conditions may vary from below freezing to tropical temperatures accompanied by high humidity. Cold reduces human efficiency, chiefly because of the bulk of the additional clothing which is required, including gloves, and because of the effects of cold wind on exposed parts of the body. The bulk of additional clothing means that larger access openings are needed to work through, but the most serious effect is often that the use of gloves impairs the sense of touch, making it more difficult to use the hands. These factors must be considered during the design stage.

The effects of high temperature and humidity are rather different from the effects of cold. They reduce the amount of physical work which can be performed, and cause a deterioration of mental ability. Besides impairing performance and producing fatigue, extreme conditions can even lead to physical collapse. Temperatures inside a vehicle or aircraft parked in the open may be considerably higher than those outside. In a hot and humid environment it is more than ever necessary to ensure that all maintenance operations which may have to be carried out are the simplest possible, and that there is a good system of part identification, that access is easy, and that parts can be quickly detached and replaced.

Wind is another element which may have to be reckoned with in a hostile environment. Aperture coverings which stay open safely in the design laboratory may be slammed shut by high winds when the equipment is standing in the open. Devices used to lock covers must operate positively, and should show clearly whether or not they are locked in the open position.

Test equipment

A number of the features necessary for good equipment maintainability apply also to test equipment, since a failure here will also delay maintenance and repair work, thus shortening the equipment's availability. Test equipment instruments must be easy to see when in use, and operability must be good (see Chapter 7). In addition it must be possible for the test gear to be easily and safely positioned when in use, within comfortable manual and visual range. Test indicators should be as simple as possible. For instance, if only a "Satisfactory—Not Satisfactory" indication is needed, the use of a lamp or buzzer is better than a meter. If a gauge or meter must be read, the dial should be as simple as possible and not capable of being read to a greater accuracy than the test requires. In more sophisticated types of special test equipment it may be desirable to incorporate some form of quick check to show that the equipment is functioning normally. Accessories, such as leads, should be permanently attached, and there must be stowage space for all loose ancillary items so that it can be seen at a glance which, if any, are missing.

The weight, portability, robustness, reliability, and even the maintainability of test equipment must be considered with reference to the skill and training of the staff who will use it, and to the environment in which it will be used. In addition, the variety of test equipment required should be kept to a minimum.

One of the most important contributions to good maintainability is a well-planned maintenance and repair manual, since this will often be the only available guide to the way the equipment works, and if information is not clearly presented, lengthy delays can result. It must be written with its purpose, and the conditions in which it will be used, firmly in mind. Thus constant page turning to refer to diagrams or tables of values must be avoided, and some guide to step by step fault tracing should be included. The manner in which the equipment operates must be clearly described and well illustrated, with simple cross-referencing between theoretical and practical working diagrams. One of the most valuable features in a maintenance and repair manual is a list of the failures which experience has shown are most likely to occur, together with their symptoms and the method of repair; such information can often save a great deal of time.

Maintenance must never involve "poking about" inside an equipment, moving connecting cables, displacing parts, and perhaps even undoing or unsoldering parts. Experience has shown that such attempts to see if failures can be anticipated does far more harm than good, since interference with the working parts is likely to introduce an element of unreliability by creating conditions leading to failure. The maintenance and repair manual should lay down maintenance routines, and no maintenance in excess of these routines must be attempted.

A useful concept in considering maintainability is to estimate the total repair and maintenance man-hours needed in a given period. Such an estimate involves employing a representative technician with a specified education, training, and experience, and timing him in the performance of the various maintenance tasks. From this the total maintenance man-hours in a given period can be calculated. The total repair man-hours can be derived in a similar way, but it is necessary to time the carrying out of various repairs and to make a reasonably accurate estimate of the failure rates of various parts, so as to establish how often each repair will be needed within the given period.

The MTBF is a measure of the likelihood that an equipment will break down in a given period, but it is also necessary to know how long it will be out of service for maintenance and repair. This time is often designated MTTR (mean time to repair). Only by taking both these factors into account can the user estimate for how long within a given period an equipment is likely to be available, and how serious the effects of non-availability during maintenance and breakdown are likely to be. He will also be able to estimate what, if any, standby equipment is necessary to obtain a certain availability of service, and avoid both the unnecessary expense of standby equipment which is not required, and the extra costs due to the effects of the equipment being out of action for longer periods than might otherwise have been anticipated.

CHAPTER 9

Reporting Failures

Serious accidents in the air, at sea, or on the railways are the subject of detailed examination to establish their cause. If an element of unreliability is involved, the investigation will reveal it, so that a suitable redesign can prevent a similar failure in the future. This process closes what is usually an open loop between the designer and the user. Normally the designer, having developed an equipment or machine to meet the user's needs in given working conditions, receives no regular and accurate information on how his design is performing. Such reports would, however, enable him to improve subsequent designs, with a resulting increase in reliability.

Reporting failures in use might well establish the shortcomings in one particular part, and so indicate how a minor and inexpensive modification could produce a substantial improvement in reliability. While it is true that information on a particular failure will usually get back to the manufacturer eventually if the failure is sufficiently widespread, he may in the meanwhile have produced and sold a substantial number of equipments which will be liable to the same failure. Furthermore, he may never hear about other failures which might otherwise have been eliminated in later models.

To improve reliability there is therefore a need to report back failures. Failure reporting should not be confined to those defects which actually lead to a breakdown. Unduly rapid deterioration in a part which may be revealed by normal maintenance procedure is clearly a failure of the part, even though it may be replaced while the equipment is still functioning normally and so not have caused a breakdown. The presence of such parts is obviously a potential cause of equipment unreliability, and in addition the need to replace them at relatively frequent intervals increases maintenance time and reduces the equipment's availability.

The most important aspects of failure reporting are accuracy and speed. If the designer receives wrong information about the nature of the failure, reporting it may do more harm than good. This means that the nature of the failure must be accurately diagnosed, the part which has failed must be identified, and the information incorporated in a report in a way which will convey the information clearly. The report should make it plain which are the facts and which are opinions or deductions about the cause of failure. If a special form has to be completed when reporting failures, it should not call for too much information, otherwise the accuracy of the details provided will depend unduly on the technical and descriptive ability of the writer and the time he has available to complete the report.

In reporting failures, or in devising forms or procedures whereby this can be done, the aim must always be to ensure that the information given is accurate, and this will usually mean that it is also necessarily limited. A manufacturer may not contemplate a redesign unless he has a number of reports all indicating the same failure. The information available in these reports collectively may be sufficient to indicate how the failure can be eliminated. If the information is not sufficient for this, it will provide a starting point for the designer to make his own investigations of the failure, and on which to base a redesign.

Reporting failures is a two-way business. The user must appreciate that accurate reports are the most important method of improving reliability of an equipment or machine after production; the designer must realize that failure reports afford the chief source of data on which the reliability of future designs can be improved. The user must ensure that the facts of a failure are reported accurately and quickly; the designer must not expect too much information to be provided, and must be prepared to find out for himself any additional information he needs to improve his design.

To sum up, however carefully a design has been conceived and manufactured, failures which reduce the reliability may nevertheless occur in particular parts when the equipment or machine is in use. If the nature of these failures is reported accurately to the designer, he can make improvements which can increase reliability.

This is particularly true when a Reliability Improvement Programme is operated during the development of an equipment. A prototype equipment or equipments are run for periods considerably in excess of the predicted MTBF (say 10–20 times). Each failure of the equipment is recorded and analysed; some of them will be random component problems (those accounted for in the MTBF prediction), others may be design related. Analysis shows up the design weaknesses, so that necessary modifications can be incorporated; therefore the MTBF should increase during the programme.

All those given the task of reporting failures should realize that it is not just a formality, but a vital procedure for improving reliability. If this is understood, reports are far more likely to be completed quickly and accurately.

CHAPTER 10

The Cost of Reliability

An important aspect of any equipment is usually its cost. Orders are normally placed with the supplier who can provide equipment or machines to the required specification at minimum initial cost, and there is no doubt in the minds of many of those who carry financial responsibility that this is the most economical way of running their business.

Nevertheless, this is very often a mistaken approach, because it takes into consideration only the initial cost and neglects the expense of keeping the equipment working satisfactorily once it has been purchased. How much maintenance and repair will cost depends on the reliability of the equipment; the more reliable an equipment is the cheaper will be the maintenance and repair bill, but the initial cost will tend to be higher.

Three separate cost factors are involved—the cost of design (including development), the cost of production, and the cost of repair and maintenance. As the reliability of an equipment increases, so will the cost of design and production increase, whereas the cost of repair and maintenance will go down. Design becomes more expensive because more precise assessments of the exact working conditions must be made, followed by more detailed development, possibly involving trials on prototypes, further environmental testing, etc. (see chart, pages 27 and 28).

On the production side higher reliability means better quality and therefore more expensive parts. It may be necessary to use costlier materials, to work to finer limits, and to provide additional and more elaborate test and inspection facilities. Usually more skilled and, therefore, more highly paid assemblers must be employed, and the completed equipment will in turn have to meet a tight and comprehensive test and inspection schedule.

To make an equipment more reliable is therefore bound to increase its initial cost. This increase can, however, be more, sometimes very much more, than offset by economies in maintenance and repair costs.

When an equipment fails there is a loss of production or of service, and often of goodwill, all of which involves some form of direct or indirect financial loss. If the results of a failure are likely to be serious, then it may be necessary to provide one or even more spare equipments as replacements. Clearly the lower the reliability the greater will be the number of equipments or machines which are out of action at any given time, and therefore the higher the number of replacements which must be provided. What is more, if the equipment is part of a system, entire spare systems may be required. Take, for instance, a weapons system such as that in a fighter aircraft. With the Service requirement that a specified number of fighters must be operational at one time, complete spare aircraft must be provided to replace those grounded by equipment failure. The need to maintain a specified service in, say, electricity or in a telephone network will also require the provision of standby equipment or systems.

Besides the financial loss caused by an equipment failure, there is the cost of repair, which is far more than just the cost of the work and material involved, since it must also take into account the expense of educating and training the necessary skilled men, the cost of test and repair apparatus and installations, and the cost of spares. The less reliable an equipment is, the more repair work is needed, and the greater the number of men and the quantity of test apparatus and of spares which must be available.

With modern complex equipment, highly skilled technicians are required for repair work and training them can be a long and expensive process. No explanation is needed of the fact that more repair work will require more and larger workshop installations, and also more apparatus for carrying out the repairs, such as tools and test gear. But all the expenses involved in the provision of spares may not be quite so obvious. Quite apart from the cost of producing spares, they perform no useful function until they are used, in contrast to an actual equipment. The total value of the stock of spares thus represents money lying idle, which, if invested or used to pay for capital goods, could produce an income. The potential income which could be derived from the money value of the stock or spares must therefore be added to the price of the spares to represent their true cost. Furthermore, spares must be ordered, transported, perhaps tested, issued, and accounted for. All this requires administrative staff and facilities.

Maintenance costs may not only mean that it costs more to keep an item in working order. They may also add to the initial cost because something must be included in the selling price to cover the average amount of repair work which it is estimated will be necessary while the item is under guarantee. The less reliable the equipment, the larger this amount will be.

The manufacture may also be affected in another way. The rate at which an equipment can be produced may be lessened by the need to divert parts, which could be incorporated in complete equipments, for use as spares.

More often than not a buyer is interested only in the initial cost of an equipment, instead of in its life cycle cost, or total cost, namely how much it will cost him to buy the equipment and to keep it functioning normally throughout its working life. By paying more initially, to obtain a more reliable equipment, maintenance costs can be made lower and the total cost reduced. This is shown in Fig. 10.1, which illustrates how production and design costs rise as reliability increases, but at the same time maintenance and repair costs fall. These three added together represent the total cost of the equipment. If the total cost is plotted it first falls with increasing reliability, and then rises again, showing that at a certain reliability the total cost is a minimum, as represented by the dotted lines in Fig. 10.1.

Note particularly that when the initial cost, represented by the production and design costs, is lowest, the total cost, which represents the total which the buyer will have to spend on the equipment during its working lifetime, is higher.

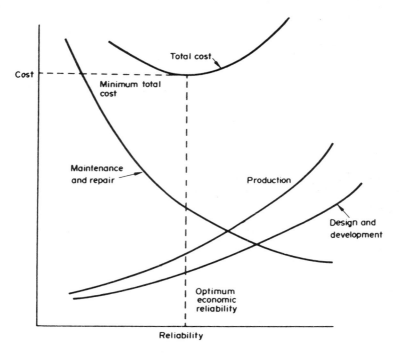

FIG. 10.1. Variation of total cost with reliability.

Optimum economic reliability, which is the degree of reliability for which the total cost is a minimum, is difficult to realize in practice because it involves estimating how production, design and maintenance costs will vary for a given reliability, and this can only be done with any accuracy if there is a considerable amount of previous experience to draw on. It may therefore be extremely difficult to make such estimates for new types of product, or for products involving new design features, new techniques, or new production methods.

Minimum total cost is of course not always the governing factor in deciding the degree of reliability of an equipment. For instance, it may be necessary to reduce reliability to the minimum acceptable to the buyer whose main consideration is an initial cost within the price range he can afford, even though he realizes that higher performance and reliability may be obtainable at a higher price. Many consumer goods fall into this category, which is usually highly competitive. Nevertheless, the reliability achieved by such equipment, for instance cars, television sets, or refrigerators, is surprisingly high when viewed against the complexity of the item and its initial cost.

As an example, the following is an analysis of car maintenance carried out in the U.S.A. some years ago. Although the costs today could be higher, the principle is the same.

Part	Contribution to breakdown probability	Maintenance action	Maintenance cost (dollars)
Ignition	0.03	Major tune up	25
Tyres	0.05	Replace	100
Engine	0.08	Overhaul	300

In terms of dollars:

Ignition   0.03/$25   = 0.0012 per $
Tyres   0.05/$100 = 0.0005 per $
Engine   0.08/$300 = 0.00027 per $

This shows that reliability improvement is preferable by maintenance on ignition and tyres rather than engine overhaul. By spending only $125 on the first two items, the same reduction in breakdown probability can be accomplished as by spending $300 on the last one.

A case where total cost is not the most important factor in determining reliability is when the buyer stipulates a high degree of reliability in the light of what he knows of the exacting conditions of use and of the extreme results of failure. These are the circumstances which increasingly often govern the purchase of equipment to be used, for example, by the Armed Forces, by civil airlines or in space. To specify a particularly high reliability requires lengthy and careful consideration of a large number of factors, some of them conflicting, and usually requires very close co-operation between the user and the manufacturer, involving perhaps an initial feasibility study to determine the best design which can be realized in practice, as well as extensive field trials of prototypes to ensure that the chosen design meets the equipment specification and will have the expected degree of reliability.

Finally there is the case where the very highest reliability possible must be obtained almost irrespective of cost. This condition must be met, for instance, in amplifying equipment in an undersea cable, because of the extremely high cost of raising the faulty length of cable for repair. Another example which may be quoted is in space research, where a failure may endanger an astronaut's life, or render useless a project costing millions of pounds. It is in fields such as these, especially those concerned with space projects, that the frontiers of reliability knowledge are being advanced, with results which will eventually help to raise the reliability of many other products.

The relationship between reliability and cost illustrates the well-known rule that you only get what you pay for. By not paying for reliability built into the product a much greater cost may eventually be incurred by the necessity for frequent repair.

CHAPTER 11

Some Useful Reliability Definitions

N.B. There are two internationally agreed specifications on reliability definitions: they are:

(1) IEC STD 271 and 271A "List of basic terms, definitions and related mathematics for reliability".
(2) MIL STD-721C "Definitions of terms for reliability and maintainability".

Some useful definitions for the reader are:

Accelerated test: A test in which the applied stress level is chosen to exceed that stated in the reference conditions, in order to shorten the time required to observe the stress response of the item or magnify the response in a given time. To be valid, an accelerated test must not alter the basic modes and/or mechanisms of failure or their relative prevalence.

Availability: The probability that the item (system) is operating satisfactorily at any point in time when used under stated conditions, where the total time includes operating time, active repair time and administrative time.

Burn-in: The operation of items prior to their ultimate application intended to stabilize their characteristics and to identify early failures.

Conformity: The ability of an item to meet its stated performance and/or characteristic requirements, the assessment of which does not depend essentially on the passage of time.

Corrective action: A documented design process, procedure, or materials change implemented and validated to correct the cause of the failure or design deficiency.

Cut set: Set of elements whose failure will cause the system to fail.

Minimal cut set: Set of elements which has no proper subset whose failure alone will cause the system to fail.

Debugging: The operation of an equipment or complex item prior to use to detect and replace parts that are defective or expected to fail and to correct errors in fabrication or assembly.

Degradation: A gradual impairment in ability to perform the required function.

Derating: The intentional reduction of stress/strength ratio in the application of an item, usually for the purpose of reducing the occurrence of stress-related failures.

Downtime: The total time during which the system is not in an acceptable operating condition.

System effectiveness: The probability the system can successfully meet an operational demand for a given time under specified conditions.

Failure: The termination of the ability of an item to perform its required function.

Failure, catastrophic: Failures that are both sudden and complete.

Failure, complete: Failures resulting from deviations in characteristic(s) beyond specified limits such as to cause complete lack of the required function. Note that limits referred to in this category are special limits specified for this purpose.

Failure, degradation: Failures that are both gradual and partial.

Failure, gradual: Failures that could be anticipated by prior examination.

Failure, misuse: Failures attributable to the application of stresses beyond the stated capabilities of the item.

Failure mode: The mechanism through which the failure occurs, that is short, open, fracture, excessive wear, etc. Also, the effect by which a failure is observed, e.g. an open or short circuit condition, or a gain change.

Failure, partial: Failures resulting from deviations in characteristic(s) beyond specified limits but not such as to cause complete lack of the required function.

Failure, random: Any failure whose cause and/or mechanism make its time of occurrence unpredictable.

Failure, secondary: Failure of an item caused either directly or indirectly by the failure of another item.

Failure, sudden: Failures that could not be anticipated by prior examination.

Failure, wear-out: A failure that occurs as a result of deterioration processes or mechanical wear and whose probability of occurrence increases with time.

Failure density: At any point in the life of an item, the incremental change in the number of failures per associated incremental change in time.

Failure mechanism: Physical, chemical, or other process resulting in a failure.

Constant failure rate: That period during which failures of some items occur at an approximately uniform rate.

Early failure period: That period during which the failure rate of some items is decreasing rapidly.

Failure rate (hazard): At a particular time, the rate of change of the number of items that have failed divided by the number of items surviving.

Failure rate, assessed: The failure rate of an item determined within stated confidence limits from the observed failure rates of nominally identical items. Note that, alternatively, point estimates may be used, the basis of which must be defined.

Infant mortality: The initial phase in the lifetime of a population of a particular component when failures occur as a result of latent defects, manufacturing errors, etc.

Instantaneous failure rate: At a particular time, the rate of change of the number of items that have failed divided by the number of items surviving.

Item (in reliability): An all-inclusive term to denote any level of hardware assembly, i.e. system, segment of a system, subsystem, equipment, component, part, etc. Note that the item includes items, population of items, sample, etc., where the context of its use so justifies.

Failure effect: The consequence(s) a failure mode has on the operation, function, or status of an item. Failure effects may be classified as local effects, next-higher effects and end-item effects.

Maintainability: The probability that when maintenance action is initiated under stated conditions, a failed system will be restored to operable condition within a specified total downtime.

Mean life: The arithmetic mean of the times to failure of a group of nominally identical items.

Mean life, assessed: The mean life of a non-repaired item determined within stated confidence limits from the observed mean life of nominally identical items. Alternatively, point estimates may be used, the basis of which must be defined.

Mean life, predicted (for non-repaired items): The mean life of an item computed from its design considerations and, where appropriate, from the reliability of its parts in the intended conditions of use.

Mean time between failures (for repairable items): The product of the number of items and their operating time divided by the total estimated number of failures.

Mean time between failures, assessed: The mean time between failures of a repairable item determined within stated confidence limits from the observed mean time between failures of nominally identical items. Alternatively, point estimates may be used, the basis of which must be defined.

Mean time between failures, observed (for repairable items): For a stated period in the life of an item, the mean value of the lengths of observed times between consecutive failures under stated stress conditions. The criteria for what constitutes a failure should be stated.

Mean time between failures, predicted: The mean time between failures of a repairable complex item computed from its design considerations and from the failure rates of its parts under the intended conditions of use. The method used for the calculation, the basis of the extrapolation, and the time-stress conditions should be stated.

Mean time to failure (for non-repairable items): The total operating time of a number of items divided by the total number of failures.

Mean time to failure, assessed: The mean time to failure of a non-repairable item determined within stated confidence limits from the observed mean time to failure of nominally identical items. Alternatively, point estimates may be used, the basis of which must be defined.

Mean time to failure, observed (for non-repairable items): For truncated tests for a particular period, the total cumulative observed time on a population divided by the total number of failures during the period under stated stress conditions. (1) Cumulative observed time is a product of units and time or the sum of these products. (2) The criteria for what constitutes a failure should be stated.

Mean time to failure, predicted: The mean time to failure of a non-repairable complex item computed from its design consideration and from the failure rates of its parts for the intended conditions of use. The method used for the calculation, the basis of the extrapolation, and the time-stress conditions should be stated.

MTBF: The product of the number of items and their operating time divided by the total number of failures.

MTTF: The total operating time of a number of items divided by the total number of failures.

MTTR: The total time taken to repair or replace an item.

Mean time to first failure: Time interval from the system initiation to the first system failure.

Operational readiness: The probability that, at any point in time, the system is either operating or ready to be placed in operation on demand when used under stated conditions.

Redundancy: The existence of more than one means of performing a given function.

Redundancy, active: That redundancy wherein all redundant items are operating simultaneously rather than being switched on when needed.

Partial redundancy: That redundancy wherein two or more redundant items are required to perform the function.

Redundancy, standby: That redundancy wherein the alternative means of performing the function is inoperative until needed, and is switched on upon failure of the primary means of performing the function.

Repairability: A measure of the speed with which a faulty item can be repaired or replaced and a check made that the equipment is functioning normally again.

Reliability: (1) The ability of an item to perform a required function under stated conditions for a stated period of time. (2) The characteristic of an item expressed by the probability that it will perform a required function under stated conditions for a stated period of time. Definition (2) is most commonly used in engineering applications. In any case, where confusion may arise, specify the definition being used.

Reliability, assessed: The reliability of an item determined within stated confidence limits from tests or failure data on nominally identical items. The source of the data should be stated. Results can only be accumulated (combined) when all the conditions are similar. Alternatively, point estimates may be used, the basis of which must be defined.

Reliability, extrapolated: Extension by a defined extrapolation or interpolation of the assessed reliability for durations or stress conditions different from those applying to the conditions of the assessed reliability.

Reliability, predicted: The reliability of an equipment computed from its design considerations and from the reliability of its parts in the intended conditions of use.

Screening test: A test or combination of tests intended to remove unsatisfactory items or those likely to exhibit early failures.

Serviceability: The degree of ease or difficulty with which a system can be repaired or maintained.

Step stress test: A test consisting of several stresses applied sequentially for periods of equal duration to a sample. During each period a stated stress level is applied and the stress level is increased from one step to the next.

Tie set: Set of elements whose functioning will ensure the system success.

Minimal tie set: Set of elements which has no proper subset whose functioning alone would ensure the system success.

Cycle time: The time interval from one system failure to the next system failure. The cycle time is the sum of an uptime and a downtime.

Operating time: The period during which the item (system) is operating in a manner acceptable to the operator.

Uptime: The total time during which the system is in acceptable operating condition.

Useful life: The length of time an item operates with an acceptable failure rate.

Note See also the list of specialized mathematical terms on page 15—The language of reliability statisticians.

References for Further Reading

1. K. B. Klaassen and J. C. L. Van Peppen, *System Reliability—Concepts and Applications*, Edward Arnold, London, 1989.
2. B. S. Dhillon, *Mechanical Reliability: Theory, Models and Applications*, American Institute of Aeronautics and Astronautics, Inc., U.S.A., 1988.
3. E. R. Hnatek, *Integrated Circuit Quality and Reliability*, Marcel Dekker, Inc., U.S.A., 1987.
4. E. A. Amerasekera and D. S. Campbell, *Failure Mechanisms in Semiconductor Devices*, John Wiley and Sons, Chichester, 1987.
5. B. S. Dhillon, *Human Reliability with Human Factors*, Pergamon Press, Oxford, 1986.
6. H. Ascher and H. Feingold, *Repairable Systems Reliability—Modeling, Inference, Misconceptions and Their Causes*, Marcel Dekker, Inc., U.S.A., 1984.
7. F. Jensen and N. E. Petersen, *Burn-In—An Engineering Approach to the Design and Analysis of Burn-in Procedures*, John Wiley and Sons, Chichester, 1982.
8. R. H. Caplen, *A Practical Approach to Reliability*, Business Books Limited, London.
9. British Standard 5760, *Reliability of Systems, Equipment and Components*.
10. G. W. A. Dummer and N. B. Griffin, *Electronics Reliability—Calculation and Design*, Pergamon Press, Oxford, 1966.